Australian
GEOGRAPHIC

Australia's Amazing
Dinosaurs

Australia's Amazing Dinosaurs

First published in 2019. Reprinted in 2023.
Australian Geographic Holdings Pty Ltd
52–54 Turner Street
Redfern NSW 2016
editorial@ausgeo.com.au
australiangeographic.com.au

Australian GEOGRAPHIC

Australian Geographic contributes
100% of its profits to the Australian
Geographic Society, including its
conservation and sustainability programs.

ALL PROFITS DONATED — AUSTRALIAN GEOGRAPHIC SOCIETY

Creative Director: Mike Ellott
Design Intern: Sophie Freeman
Dinosaur Consultant: Timothy Frauenfelder
Managing Editor: Katrina O'Brien
Managing Commercial Editor: Lauren Smith
Assistant Commercial Editor: Rebecca Cotton
Managing Director: Jo Runciman
Editor-in-Chief: Chrissie Goldrick

Printed in China by RR Donnelley

NATIONAL LIBRARY OF AUSTRALIA
A catalogue record for this
book is available from the
National Library of Australia

FSC
www.fsc.org
MIX
Paper | Supporting
responsible forestry
FSC® C020056

Contents

Introduction:

Ancient reptiles

DINOSAURS ARE land-based reptiles that lived between 227 and 66 million years ago. Other prehistoric reptiles, like the pterosaurs that flew in the air and the pliosaurs that swam in the ocean, aren't actually dinosaurs.

Dinosaurs can be divided into two groups. There are the saurischian ('lizard-hipped') dinosaurs that include giant long-necked sauropods (such as *Brontosaurus* and *Diamantinasaurus*) and two-legged **carnivorous** theropods (such as *Tyrannosaurus* and *Australovenator*).

The other group are the ornithischian ('bird-hipped') dinosaurs that include frilled and armoured species (such as *Triceratops* and *Minmi*) and ornithopods – **herbivorous** herd-living dinosaurs (such as *Iguanadon* and *Muttaburrasaurus*).

All animals that are descended from the last common ancestor of both these groups can be classified as dinosaurs. This means that all bird species – even the ones we know today – are technically dinosaurs!

Fossil evidence

EVEN THOUGH they lived a long time ago, we know about dinosaurs because of their fossils. There are two types of fossils. Body fossils are the fossils of bones that got covered up by a sediment such as rock or sand. In most cases, minerals seep into the bones, slowly replacing the bone but keeping its shape, as mud and sand harden into rock around it.

The rock protects the fossil, and when **palaeontologists** find it they very carefully dig it out of the dirt. Fossils can break easily, so the scientists have to be very gentle.

From fossils, even tiny ones, palaeontologists are often able to get an understanding of what a dinosaur looked like, what it ate or even what the environment it lived in was like.

The second kind of fossil is a trace fossil. This is where a sign of animal activity has been preserved, like the dinosaur footprints at Lark Quarry in outback Queensland.

An ornithopod footprint from Broome, Western Australia.

THAT'S AMAZING!
Australian dinosaur body fossils are often only fragments or pieces, such as this theropod tooth.

Ozraptor

Ozraptor subotaii

How to say it:

"oz-rap-tor"

IN 1966 A SHINBONE was found by four 12-year-old boys near Geraldton in Western Australia. When a model of the bone was sent to England, experts first suspected that the bone belonged to a turtle. *Ozraptor* is now thought to have been a speedy meat-eater – and is the oldest of all the dinosaurs discovered in Australia.

THAT'S AMAZING!

The partial shinbone found in 1966 is still the only fossil evidence we have for this dinosaur.

Age: Mid Jurassic (170 million years old)
Type: Ceratosaur
Region: Western Australia
Diet: Carnivore
Length: 2–3m **Height:** Unknown
Weight: 100kg

Rhoetosaurus

Rhoetosaurus brownei

How to say it:
"ree-toe-sore-us"

FACT:
This was the first big dinosaur to be found in Australia, and the **stockmen** who found the first fossils thought they were the remains of an escaped circus elephant!

THE FIRST FOSSILS of *Rhoetosaurus* were found in 1924 in Roma, Queensland, and more of the same skeleton was uncovered in 1976. It is named after the mythical Greek giant Rhoetos and would have used its long neck to reach up to plants like **conifers**. *Rhoetosaurus* would have been a fairly slow dinosaur, only capable of moving at about 15 km/h, although it might not have moved very much at all.

Age: Mid Jurassic (167 million years old)
Type: Sauropod
Region: Queensland
Diet: Herbivore
Length: 14m **Height:** 3.5m
Weight: 20 tonnes

Dinosaur family tree

Australia's dinosaurs come from many different branches of the dinosaur family tree and are related to famous species from other parts of the world.

ORNITHISCHIANS
Bird-hipped dinosaurs

Ornithopods
Two-legged herbivores
Hypsilophodon, Parasaurolophus, Muttaburrasaurus, Leaellynasaura, Qantassaurus, Atlascopcosaurus, Weewarrasaurus, Galleonosaurus, Fulgurotherium, Fostoria

Dinosaur Group
Group Characteristics
Examples of group members
Australian species

Pachycephalosaurs
Dome-headed and two-legged
Pachycephalosaurus, Stegoceras

Thyreophorans
Armoured dinosaurs

Ceratopsians
Four-legged, frilled and horned
Triceratops, Centrosaurus, Serendipaceratops

Stegosaurs
Four-legged herbivore with plates and spikes
Stegosaurus

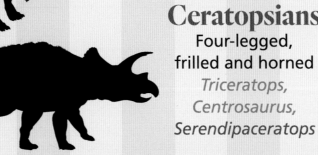

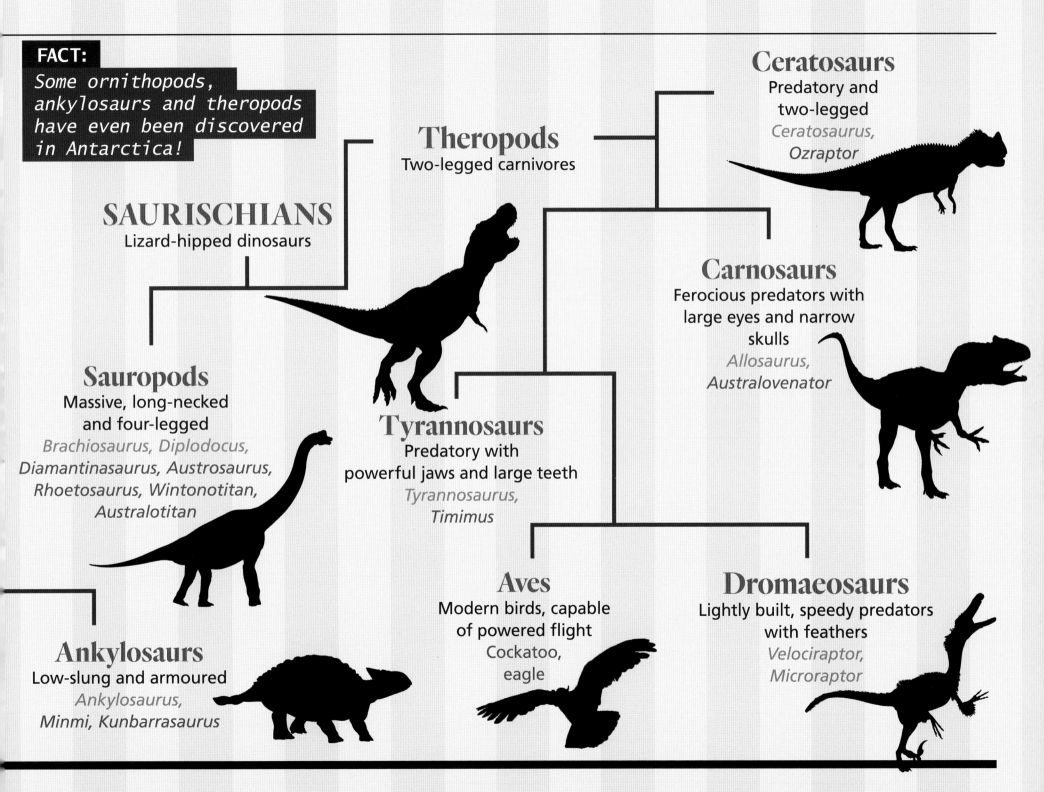

SAURISCHIANS
Lizard-hipped dinosaurs

Theropods
Two-legged carnivores

Ceratosaurs
Predatory and two-legged
Ceratosaurus, Ozraptor

Carnosaurs
Ferocious predators with large eyes and narrow skulls
Allosaurus, Australovenator

Sauropods
Massive, long-necked and four-legged
Brachiosaurus, Diplodocus, Diamantinasaurus, Austrosaurus, Rhoetosaurus, Wintonotitan, Australotitan

Tyrannosaurs
Predatory with powerful jaws and large teeth
Tyrannosaurus, Timimus

Ankylosaurs
Low-slung and armoured
Ankylosaurus, Minmi, Kunbarrasaurus

Aves
Modern birds, capable of powered flight
Cockatoo, eagle

Dromaeosaurs
Lightly built, speedy predators with feathers
Velociraptor, Microraptor

In focus:

Minmi

Minmi paravertebra

How to say it:

"min-my"

WHEN SOME REMAINS of this stocky species were first found in the 1960s, it was instantly recognised as an armoured ankylosaur type of dinosaur. It had bony, protective plates (scutes) on its back as well as on its stomach. This armour might have protected it from carnivorous raptors.

This was the first known ankylosaurian from Australia. When a second, nearly complete specimen of an ankylosaur was found in 1989, it was believed to be a *Minmi* specimen, but in 2015 it was identified as a new species of ankylosaur – a *Kunbarrasaurus*.

Age: Early Cretaceous (125 million years old)
Type: Ankylosaur
Region: Queensland
Diet: Herbivore
Length: 3m
Height: 1m **Weight:** 100kg

FACT :

Minmi is thought to have had a particularly small brain.

Galleonosaurus

Galleonosaurus dorisae

How to say it:

"gal-ee-uhn-oh-sore-us"

ONE OF THE most recently identified of Australia's ornithopods is *Galleonosaurus dorisae*, the fifth of its kind named from Victoria. The dinosaur was given its name because the shape of its jaw resembles a type of ship known as a galleon.

Five different upper-jaw bones of *Galleonosaurus* were found at Flat Rocks, a fossil site near Inverloch, Victoria. Flat Rocks would have been a lush, forested area 125 million years ago, with lots of plants for these agile herbivores to feed on. *Galleonosaurus* lived in the same time and region as *Qantassaurus*, but the shape of each species' jaw shows us that they fed on different types of plants.

Age: Early Cretaceous (125 million years old)
Type: Ornithopod Region: Victoria
Diet: Herbivore Length: 2.4m
Height: 1.4m Weight: 25kg

THAT'S AMAZING!

This dinosaur is likely to have been around the same size as a wallaby.

Dinosaurs by size and time

Discover just how big the Australian dinosaurs were, and see when they walked this land.

110–106 mya
Timimus

THAT'S AMAZING!

Rhoetosaurus *weighed about 20 tonnes. The heaviest known dinosaur in the world,* Argentinosaurus, *could have weighed up to 90 tonnes.*

167–165 mya
Rhoetosaurus

125 mya
Galleonosaurus

144–125 mya
Minmi

125–120 mya
Qantassaurus

MYA: million years ago

225 mya
First mammals

170–168 mya
Ozraptor

160 mya
Oldest flowering plant and bird fossils

227 mya
First dinosaurs

Triassic Period

(251–201 mya)

Jurassic Period

(201–145 mya)

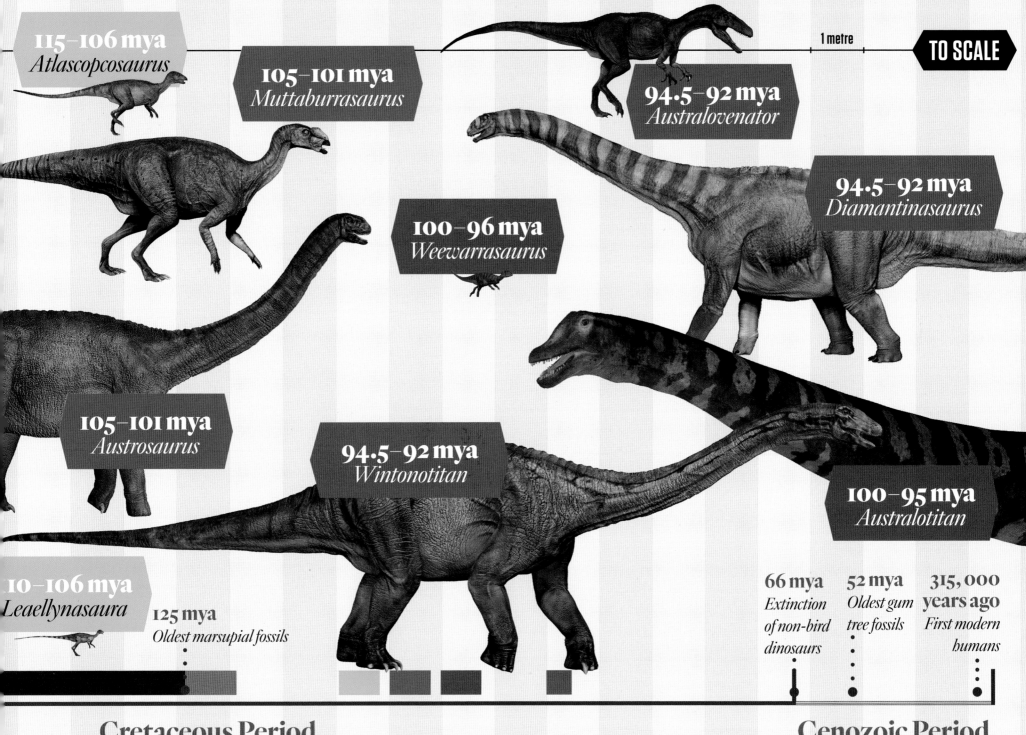

115–106 mya
Atlascopcosaurus

105–101 mya
Muttaburrasaurus

94·5–92 mya
Australovenator

1 metre

TO SCALE

94·5–92 mya
Diamantinasaurus

100–96 mya
Weewarrasaurus

105–101 mya
Austrosaurus

94·5–92 mya
Wintonotitan

100–95 mya
Australotitan

110–106 mya
Leaellynasaura

125 mya
Oldest marsupial fossils

66 mya
Extinction of non-bird dinosaurs

52 mya
Oldest gum tree fossils

315,000 years ago
First modern humans

Cretaceous Period
(145–66 mya)

Cenozoic Period
(66 mya – today)

AUSTRALOTITAN ART: VLAD KONSTANTINOV © EROMANGA NATURAL HISTORY MUSEUM

In focus:

Qantassaurus

Qantassaurus intrepidus

QANTASSAURUS WAS a small dinosaur – about the size of a kangaroo – that was discovered by Patricia Vickers-Rich and Tom Rich (who also named *Leaellynasaura* and *Timimus*) in Inverloch, Victoria, in 1994. This dinosaur was a herbivore that used its muscular cheeks and grinding teeth to eat a wide range of low-growing plants such as ferns.

Age: Early Cretaceous (120 million years old)
Type: Ornithopod **Region:** Victoria
Diet: Herbivore **Length:** 1.8m
Height: 1m **Weight:** 45kg

How to say it:
"kwant-uh-sore-us"

THAT'S AMAZING!
This species had large eyes compared with most dinosaurs, which would have helped it see better during the night.

Atlascopcosaurus

Atlascopcosaurus loadsi

ONLY A FEW BONES from the head of this dinosaur have been discovered, back in 1989, but it's enough to reveal that it was a member of the ornithopods – a group of two-legged herbivores. Most feathered species found so far have been carnivorous dinosaurs, but there are a few hints that herbivores like *Atlascopcosaurus* may have been feathered as well.

How to say it:
"at-lass-cop-coe-sore-us"

FACT:

Atlascopcosaurus *is named for a mining company called Atlas Copco, who sponsored the fossil dig.*

Age: Early Cretaceous (115 million years old)
Type: Ornithopod
Region: Victoria
Diet: Herbivore
Length: 3m Height: 1m
Weight: 125kg

What dinosaurs looked like

Did dinosaurs have scaly skin or feathers? Were they brightly coloured, or dull and drab? We're still learning the answers to both these questions.

WHAT COLOUR dinosaurs were has always been a bit of a mystery. Even when an impression of dinosaur skin was found (usually a scaly or pebbly texture) scientists had no way of knowing its colour. Some thought dinosaurs would have been dull to help them blend into their environment. If you think of today's big land animals – like elephants, rhinoceroses and hippopotamuses – they're usually drab grey or brown. Others believed dinosaurs would have been vivid like birds, with purple, orange, red or blue skin to help them identify one another and find mates. The discovery of feathered dinosaurs in 1996 helped solve some of the mystery. The colour of a feather (or of your hair) is determined by a protein called keratin inside it. When viewed under a high-powered microscope, certain pigments in keratin can reveal the colour of feathered, fossilised dinosaurs.

THAT'S AMAZING!
This Sinosauropteryx fossil was one of the first feathered dinosaurs analysed. Experts discovered that it was ginger coloured, with a striped tail.

Australotitan

Australotitan cooperensis

THIS GIGANTIC DINOSAUR is the biggest found in Australia so far and one of the 20 largest dinosaurs found anywhere in the world! It was uncovered in 2005 in Eromanga, Queensland, but it took 17 years for the Queensland Museum and Eromanga Natural History Museum to carefully extract its partial skeleton and study the bones to be sure it was a new, colossal titanosaurian. Nicknamed 'Cooper' as it was found near Cooper Creek, this plant-eating sauropod was formally described by Dr Scott Hocknull in 2021 and is believed to have shared a common ancestor with the smaller *Wintonotitan* and *Diamantinasaurus*.

Age: Mid Cretaceous (100 million years old)
Type: Sauropod
Region: Queensland **Diet:** Herbivore
Length: 30m **Height:** 6.5m
Weight: 63 tonnes

THAT'S AMAZING!

At least three other fossils found at Eromanga are thought to belong to as-yet-undescribed Australotitans.

How to say it:

"Oss-trah-low-tie-tan coo-per-ennsiss"

© EROMANGA NATURAL HISTORY MUSEUM.

In focus:

Timimus

Timimus hermani

WHEN THESE FOSSILS were first discovered in Dinosaur Cove in Victoria in 1991, scientists thought the dinosaur they came from was related to a group of herbivores called ornithomimosaurs, or 'bird-mimic lizards', which looked a bit like ostriches. Scientists now think *Timimus* is actually a distant relative of *Tyrannosaurus rex*!

Age: Early Cretaceous (105 million years old)
Type: Tyrannosaur
Region: Victoria **Diet:** Carnivore
Length: 3.5m **Height:** 1.5m
Weight: 4 tonnes

FACT :

Where Timimus was running around, it would have been very cold and dark in winter, so it is believed that this dinosaur hibernated through the cold seasons.

Leaellynasaura

Leaellynasaura amicagraphica

THIS SMALL plant-eater lived in southern Australia when it was partly within the Antarctic Circle, where temperatures ranged between −2 °C and 5 °C. The skull that was found at Dinosaur Cove in Victoria shows that it had a large brain and large eyes, which helped it keep watch for predators as it looked for food in the long, dark Antarctic winters. Its incredibly long tail was more than twice the length of its body – perhaps the longest tail (relative to body length) of all dinosaurs.

Age: Early Cretaceous (110 million years old)
Type: Ornithopod
Region: Victoria Diet: Herbivore
Length: 1.5m Height: 0.6m
Weight: 10kg

THAT'S AMAZING!

Leaellynasaura *was discovered in 1987 on a dig run by Patricia Vickers-Rich and Tom Rich. They named it after their daughter Leaellyn.*

How to say it:

"lee-allin-ah-sore-ah"

More Australian dinosaurs

Fostoria dhimbangunmal

Age: Late Cretaceous Region: NSW Size: 5m

Although *Fostoria* fossils were first found in 1984, it was only 33 years later, in 2019, that they were declared a new species. The fossils belonged to a small family group, or herd, of dinosaurs related to *Muttaburrasaurus*. This plant-eating dinosaur would have roamed the area around Lightning Ridge 100 million years ago.

Walgettosuchus woodwardi

Age: Early Cretaceous Region: NSW Size: Unknown

Walgettosuchus is a mysterious theropod, known from only a single fossilised tailbone found near the town of Walgett before 1909. It is tricky to figure out its size or classify it correctly from a single vertebra. Some palaeontologists believe that it and a dinosaur named *Rapator* may be the same species, but as *Rapator* is known only from a hand bone, more fossil evidence is needed.

FACT:

Fulgurotherium australe *is another small (1–1.5m) herbivorous ornithopod known from a fragment of opalised femur that was found by an opal dealer at Lightning Ridge, NSW, in the early 1900s.*

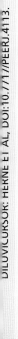

Lightning Claw

Age: Early Cretaceous **Region:** NSW **Size:** 7m

The largest known carnivorous dinosaur found in Australia, this megaraptorid was a large and slender predator. Lightning Claw didn't have as powerful a jaw as other carnivores, so it potentially used its massive claws like grappling hooks to catch prey. It would have roamed around swamps, lakes and rivers near Lightning Ridge.

Kunbarrasaurus ieversi

Age: Early Cretaceous **Region:** Qld **Size:** 3m

In 1989 a well-preserved dinosaur skeleton was found. Previously thought to be a *Minmi* specimen, in 2015, the bones were revealed to belong to a new kind of ankylosaur, *Kunbarrasaurus* ("kun-buh-ruh-sore-us"). It had a parrot-like beak and wasn't as heavily armoured as later ankylosaurs.

Diluvicursor pickeringi

Age: Early Cretaceous **Region:** Vic **Size:** 2m

The *Diluvicursor* ("die-loo-vee-kerr-sore") specimen found in 2005 – including a partial shin and foot, and a nearly complete tail – is the second-most complete Victorian ornithopod skeleton described so far. This turkey-sized dinosaur would have foraged for pine cones, ferns and other plant matter in the forested area it inhabited. It was small and strong, and it would have been a fast runner.

Savannasaurus elliottorum

Age: Late Cretaceous **Region:** Qld **Size:** 12–15m

Belonging to the herbivorous, long-necked sauropod family, *Savannasaurus* had very wide hips and stocky limbs with five toes on each foot. The 20% to 25% complete skeleton found near Winton, Queensland, tells us that this dinosaur had a neck that was likely longer than its tail, and a barrel-like torso.

THAT'S AMAZING!

The Savannasaurus *fossil has been nicknamed 'Wade' in honour of a famous Australian palaeontologist, Mary Wade.*

Muttaburrasaurus

Muttaburrasaurus langdoni

MUTTABURRASAURUS is Australia's largest ornithopod, and fossils of the species, including two skulls, have been found in both Queensland and New South Wales. This powerful herbivore could rear up on its back legs to reach tall trees and intimidate predators, although it sometimes moved about on all four legs, too.

How to say it:

"mutta-burra-sore-us"

FACT:

The unusual bulge on its snout may have contained an inflatable air sac. This may have helped it call out to other dinosaurs.

Age: Early Cretaceous (105 million years old)
Type: Ornithopod
Region: Qld & NSW
Diet: Herbivore
Length: 7–8m
Height: 2.5m
Weight: 4 tonnes

In focus:

Austrosaurus

Austrosaurus mckillopi

How to say it:
"aus-tro-sore-us"

FIRST DISCOVERED IN north-central Queensland almost 90 years ago, *Austrosaurus* was our first known sauropod from the Cretaceous period. More fossils were found in 2014 and 2015. This long-necked species was able to reach tall trees and foliage. Scientists once thought that *Austrosaurus* spent a lot of time in the water, since it was found in an area with lots of plesiosaurs (a type of marine reptile). We know now that's not correct, however, and that it was a land-based animal.

FACT :

Austrosaurus means 'southern reptile', and mckillopi *refers to the McKillop brothers, who ran the farm where the fossils were found.*

Age: Early Cretaceous (105 million years old)
Type: Sauropod
Region: Queensland
Diet: Herbivore
Length: 12–15m
Height: 3.5m
Weight: 15 tonnes

Where Australia's dinosaurs are found

Ancient fossils and dinosaur tracks have been found all over Australia!

Where is the closest spot to you?

FACT:
It's not just dinosaur fossils that have been found around Australia, but also other things such as plants, mammals and amphibians.

① Winton, Qld

More than 3300 dinosaur tracks can be seen at Lark Quarry in Winton. They date back 95 million years, and for a long time, scientists thought that they were the record of a dinosaur stampede! Scientists now suspect that the footprints were left by dinosaurs crossing a river.

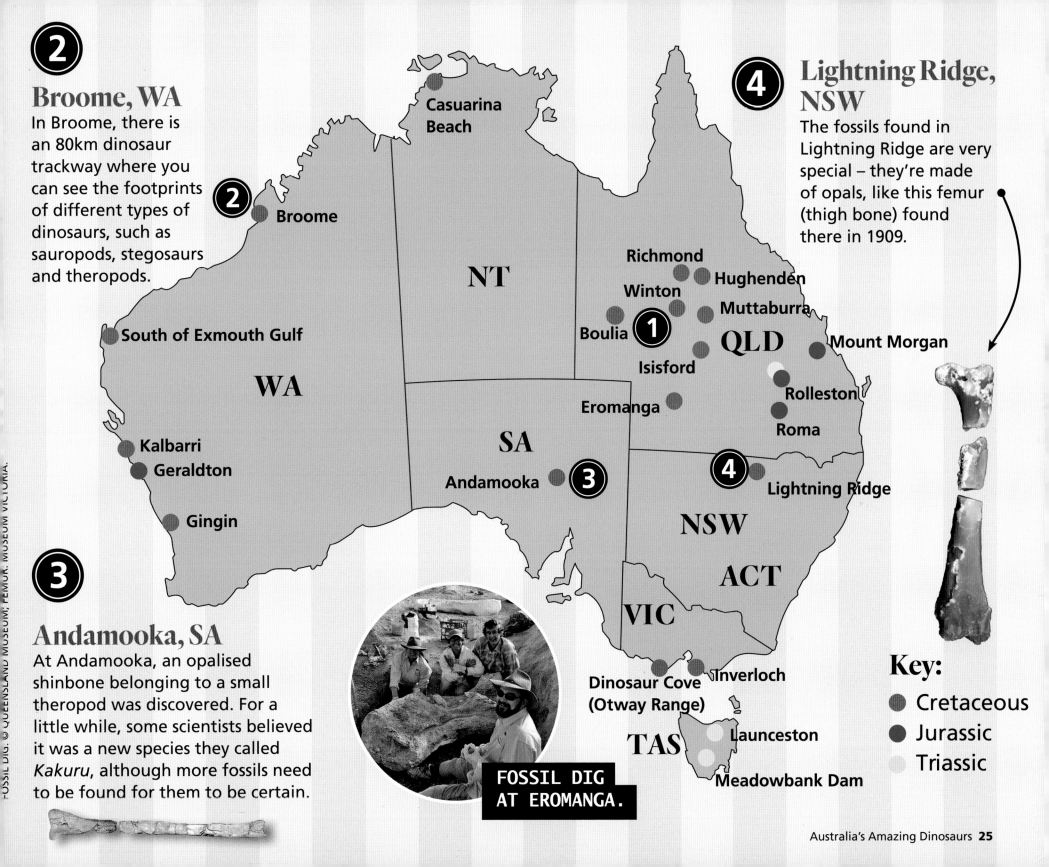

②

Broome, WA

In Broome, there is an 80km dinosaur trackway where you can see the footprints of different types of dinosaurs, such as sauropods, stegosaurs and theropods.

④ **Lightning Ridge, NSW**

The fossils found in Lightning Ridge are very special – they're made of opals, like this femur (thigh bone) found there in 1909.

③

Andamooka, SA

At Andamooka, an opalised shinbone belonging to a small theropod was discovered. For a little while, some scientists believed it was a new species they called *Kakuru*, although more fossils need to be found for them to be certain.

Casuarina Beach

NT

WA

② Broome

South of Exmouth Gulf

Kalbarri
Geraldton

Gingin

Richmond
Hughenden
Winton
Muttaburra
① Boulia
QLD
Isisford
Mount Morgan
Eromanga
Rolleston
SA
Roma
Andamooka ③
④
Lightning Ridge
NSW
ACT
VIC
Inverloch
Dinosaur Cove (Otway Range)
TAS
Launceston
Meadowbank Dam

FOSSIL DIG AT EROMANGA.

Key:

● Cretaceous
● Jurassic
Triassic

Weewarrasaurus

Weewarrasaurus pobeni

How to say it:
"wee-warr-a-sore-us"

Age: Late Cretaceous
(100 million years old)
Type: Ornithopod
Region: New South Wales
Diet: Herbivore

IN 2013, THE KEEN eyes of Mike Poben, a miner from Lightning Ridge, caught sight of two interesting-looking pieces in a bag of rough opal that had been brought up from a mine in nearby Wee Warra. Poben's suspicion that the pieces may be fossilised jawbone was confirmed by experts at the Australian Opal Centre, to which he then donated the pieces.

In 2018 a species was described from the fossils – *Weewarrasaurus pobeni*. This dog-sized ornithopod had a beak and serrated teeth for eating plants, and may have moved in a herd.

THAT'S AMAZING!
During the time Weewarrasaurus roamed, Lightning Ridge was covered by lakes and waterways and was much closer to the South Pole than it is today.

Diamantinasaurus

Diamantinasaurus matildae

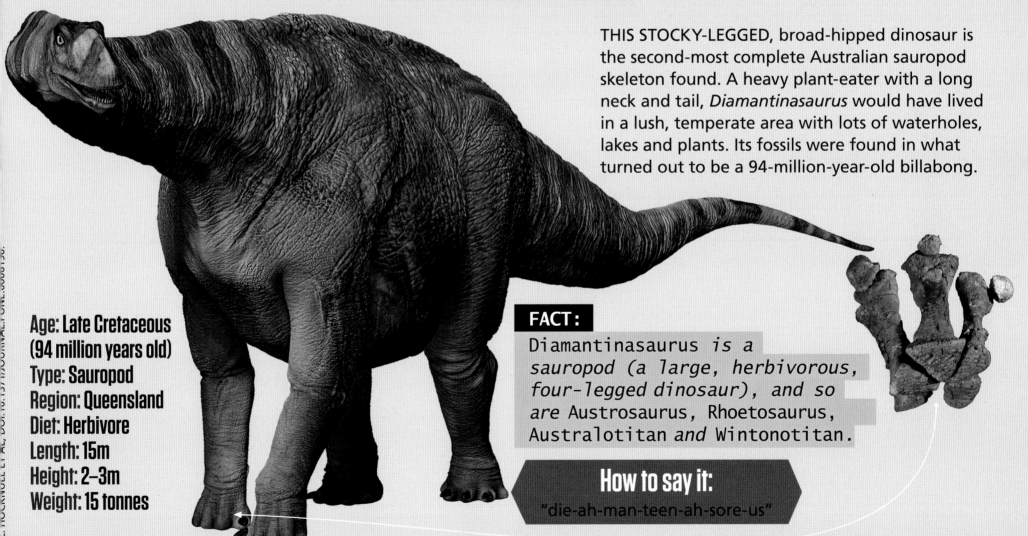

THIS STOCKY-LEGGED, broad-hipped dinosaur is the second-most complete Australian sauropod skeleton found. A heavy plant-eater with a long neck and tail, *Diamantinasaurus* would have lived in a lush, temperate area with lots of waterholes, lakes and plants. Its fossils were found in what turned out to be a 94-million-year-old billabong.

Age: Late Cretaceous
(94 million years old)
Type: Sauropod
Region: Queensland
Diet: Herbivore
Length: 15m
Height: 2–3m
Weight: 15 tonnes

FACT :

Diamantinasaurus *is a sauropod (a large, herbivorous, four-legged dinosaur), and so are* Austrosaurus, Rhoetosaurus, Australotitan *and* Wintonotitan.

How to say it:

"die-ah-man-teen-ah-sore-us"

Dinosaurs today

DINOSAURS, AS WE usually picture them, had all died out by the beginning of the Cenozoic period, 66 million years ago. When we consider that birds are descended from dinosaurs (and knowing that there are around 10,000 bird species in the world) we could say there are a lot of dinosaurs still living! Other than birds, there are no true dinosaurs left alive. Australia does have a few animal species, however, that resemble creatures that existed in the same time period as some of the dinosaurs.

Isisfordia

Isisfordia is an **extinct** reptile and a distant relative of today's crocodiles. This Cretaceous-aged croc dates to around 100 million years ago and is one of the oldest-known direct ancestors of modern crocodiles. An ambush predator estimated to grow between 1.5 and 2m in length, it might have preyed upon dinosaurs such as *Weewarrasaurus*.

THIS PAGE: JOSÉ VITOR SILVA · DOI:10.7717/PEERJ.7166/FIG-6; FOSSIL · LACHLAN HART · DOI:10.7717/PEERJ.7166/FIG-4·

Saltwater crocodile

Crocodylus porosus

Crocodiles aren't dinosaurs, but archosaurs. We know from fossil records that the first ancestors of crocodiles appeared about 240 million years ago, making them much older than any of the Australian dinosaurs we know of. These ancient ancestors weren't really too distinct from the species we're familiar with today – like the saltwater crocodile. The saltie is much younger than these ancestors, between 12 and 6 million years old, but its body hasn't **evolved** much from the first crocodilians.

FACT:
A jawbone fossil from Steropodon was found at Lightning Ridge. The area was once at the edge of a big, shallow sea. The sediments in that sea hardened into rock, preserving many plant and animal remains.

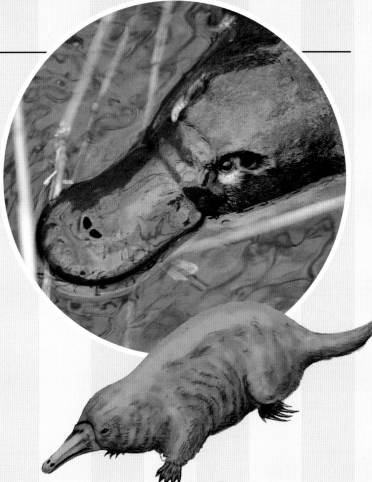

Steropodon

Steropodon galmani

The platypus has always been a bit of a mysterious creature. When European scientists first sent word of it back to England, people assumed it was a hoax! But the humble platypus has a **heritage** that goes back **aeons** – it has an ancestor, *Steropodon galmani*, that was alive about 100 million years ago during the Cretaceous period.

Wintonotitan

Wintonotitan wattsi

How to say it:
"win-ton-oh-tie-tan"

Age: Late Cretaceous (95 million years old)
Type: Sauropod **Region:** Queensland
Diet: Herbivore
Length: 16–17m **Height:** 3.5m
Weight: 20 tonnes

THE FIRST FOSSIL of this towering plant-eater was initially thought to belong to *Austrosaurus*. It was re-examined a few times in the 2000s, and researchers finally declared it a unique species, making it the fourth Australian sauropod. It is nicknamed 'Clancy' after Banjo Paterson's poem *Clancy of the Overflow*. It was found near Winton, not far from *Australovenator* and *Diamantinasaurus*.

THAT'S AMAZING!
Although the sauropods found in Queensland are large, there are also trace fossils of large sauropods near Broome in Western Australia.

Evidence of a sauropod footprint in Broome, Western Australia

In focus:

Australovenator

Australovenator wintonensis

THIS SPEEDY dinosaur had huge claws on its hands and slender-toed feet. It has been described as the cheetah of Cretaceous Australia. Compared to other Australian carnivorous dinosaurs, we have the most complete fossil record for this species. It was discovered in 2009 by palaeontologist Dr Scott Hocknull near Winton in Queensland.

FACT :

The fossil skeleton of Australovenator is nicknamed 'Banjo' after the famous Australian bush poet, Banjo Paterson.

ACTUAL SIZE!

Age: Late Cretaceous (93 million years old)
Type: Carnosaur
Region: Queensland Diet: Carnivore
Length: 6m Height: 2m Weight: 300–400kg

Glossary

Aeons
The largest division of time, covering multiple eras and millions of years.

Carnivorous
Meat-eating.

Conifers
A group of plants that flower all year and produce their seeds as cones, such as the pine, spruce or fir tree.

Evolved
Adapted from a previous generation.

Extinct
A plant or animal species that has no living members.

Herbivorous
Plant-eating.

Heritage
A link to previous generations.

Palaeontologists
Scientists who study fossils.

Protein
A molecule that is an essential part of living things.

Stockmen
People who work on farms, especially with cattle.